T0155518

SpringerBriefs in Applied Sciences and Technology

Computational Intelligence

Series editor

Janusz Kacprzyk, Systems Research Institute, Polish Academy of Sciences, Warsaw, Poland

SpringerBriefs in Computational Intelligence are a series of slim high-quality publications encompassing the entire spectrum of Computational Intelligence. Featuring compact volumes of 50–125 pages (approximately 20,000–45,000 words), Briefs are shorter than a conventional book but longer than a journal article. Thus Briefs serve as timely, concise tools for students, researchers, and professionals.

More information about this series at http://www.springer.com/series/10618

Camilo Caraveo · Fevrier Valdez
Oscar Castillo

A New Bio-inspired Optimization Algorithm Based on the Self-defense Mechanism of Plants in Nature

 Springer

Camilo Caraveo
Division of Graduate Studies
Tijuana Institute of Technology
Tijuana, Baja California, Mexico

Oscar Castillo
Division of Graduate Studies
Tijuana Institute of Technology
Tijuana, Baja California, Mexico

Fevrier Valdez
Division of Graduate Studies
Tijuana Institute of Technology
Tijuana, Baja California, Mexico

ISSN 2191-530X   ISSN 2191-5318   (electronic)
SpringerBriefs in Applied Sciences and Technology
ISSN 2625-3704   ISSN 2625-3712   (electronic)
SpringerBriefs in Computational Intelligence
ISBN 978-3-030-05550-9   ISBN 978-3-030-05551-6   (eBook)
https://doi.org/10.1007/978-3-030-05551-6

Library of Congress Control Number: 2018964225

This Springer imprint is published by the registered company Springer Nature Switzerland AG
The registered company address is: Gewerbestrasse 11, 6330 Cham, Switzerland

Preface

In this book, a new meta-heuristic algorithm is presented, and the proposal is of a new optimization algorithm bio-inspired on the self-defense mechanisms of plants in the nature. In the literature, there are many published works, where the authors scientifically demonstrate that plants have different mechanisms of self-defense (survival strategies) and these techniques are used to defend against predatory organisms, in this case herbivorous insects.

The proposed algorithm considers as a basis the predator–prey model proposed by Lotka and Volterra. The predator–prey model is formed by two nonlinear first-order differential equations that allow modeling the growth of two populations that interact with each other (prey and predator).

The main contribution of this book is the creation of a new optimization meta-heuristic that is capable of competing against the existing algorithms in the literature. In the first phase of the proposed approach, performance and stability are tested in the optimization of two sets of benchmark mathematical functions. The first set consists in traditional basic functions, the second challenge is a set of mathematical functions with a higher level of complexity called CEC-2015, and these functions were defined with the aim of creating a global competition for new meta-heuristics able to solve these functions.

The second phase of the meta-heuristics proposal was to add a fuzzy approach for dynamic parameter adaptation, the system of predator–prey equations uses four parameters (α, β, λ, δ), and the values of these variables are very important since they are in charge of maintaining a balance between the pair of equations. The final phase of this book deals with the application of Type 2 fuzzy logic for the dynamic adaptation of the parameters of the system. This time a fuzzy controller is in charge of finding the optimal values for the model parameters, and the use of this technique allows the algorithm to have a higher performance and accuracy in the exploration of the solution values.

Tijuana, Mexico

Dr. Camilo Caraveo
Prof. Fevrier Valdez
Prof. Oscar Castillo

Contents

Chapter 1
Introduction

Meta-heuristic algorithms have been very popular in recent years and are frequently used to solve optimization problems. There are many bio-inspired algorithms, such as: PSO (Particle Swarm Optimization) [1, 2], ABC (Artificial Bee Colony) [3–7], ACO (Ant Colony Optimization) [8, 9], GA (Genetic Algorithm), and GSA (Gravitational Search Algorithm). These optimization algorithms have been applied to many problems, for example: optimization of neural networks, or recently in [10–14] fuzzy logic has been used to adapt some parameters of bio-inspired algorithms, for greater performance and stability. There is also algorithm hybridization to solve multiple optimization problems such as: routing problems, function approximation, route optimization, among others. In this book a new meta-heuristic optimization algorithm is proposed. The proposed optimization algorithm is based on the self-defense mechanisms of the plants. The aim of this proposal is to explore and exploit new meta-heuristic processes that help to solve different problems and compete against traditional algorithms that have already been studied and exploited; the idea is to create an alternative that can be able to solve complex problems reducing time and computational cost. In the literature, some authors have demonstrated that plants are living organisms with biological physiological processes, such as respiration, reproduction and carrying a complex life cycle using different reproduction method such as: pollination, graft among others.

Reproduction processes are used to generate populations of new individuals that reach maturity or may repeat the same cycle, and so on, to avoid the extinction of their species and their predators [15–17]. In [18–20], the authors have published previous works using the basic ideas of this optimization algorithm applied to the optimization of benchmark mathematical functions for different number of variables and using different methods of reproduction; published results showing that the performance of the algorithm is efficient even being a new algorithm and is in the development stage and improvements [20].

The main contribution in this book is the proposal of a new optimization algorithm based on self-defense mechanisms of the plants and in addition different

C. Caraveo et al., *A New Bio-inspired Optimization Algorithm Based on the Self-defense Mechanism of Plants in Nature*, SpringerBriefs in Computational Intelligence, https://doi.org/10.1007/978-3-030-05551-6_1

methods of biological reproduction are also developed as part of this algorithm. In this case, the predator-prey model proposed by Lotka and Volterra is used as the basis for this new algorithm. The main difference of this algorithm against the prey predator model is that the proposed algorithm has an evolutionary process where plants develop coping strategies to survive from predators, and also has different methods for biological reproduction to conserve this species. In addition, each reproduction method considers different characteristics of the plant to reproduce.

We consider important to mention that in recent years the use of dynamic adaptation of parameters in meta-heuristics of optimization is a strategy that has had a very important impact in the area of bio-inspired algorithms. The results published by the authors who use this adaptation in their proposals show a very significant improvement in their results. The use of fuzzy logic in bio-inspired algorithms is an intelligent technique that enables the algorithm to make an intelligent search of values for the parameters and in combination with the algorithm select the optimal values for the solution of the problem.

In the literature, there are many works with dynamic adaptation of parameters using fuzzy logic for example: In [3–5] the authors use the bee colony algorithm using dynamic tuning of parameter to optimize a controller applied to a mobile autonomous robot. In [9] the author performs an improvement of the Ant Colony Algorithm using interval Type-2 fuzzy logic applied to the travel agent problem (TSP). In [21] the authors propose a new variant of the Particle Swarm Algorithm using FL applied to optimize mathematical functions benchmark. The works mentioned are some of those found in the literature, as we can observe the use of fuzzy controllers can be used for multiple problems, in recent years is used to improve the performance of the algorithms [22–26].

The main difference in this work is that the meta-heuristic that is being used is a totally new algorithm, the problems were previously studied by other authors using different optimization meta-heuristics such as some of the following algorithms Ant Colony [9], Bee Colony [3, 6, 8, 27], Genetic Algorithms [28], Particle Swarm, the meta-heuristics mentioned above were used to optimize the robot trajectory in order to test the performance to complex problems, some of these algorithms were able to find acceptable results and were published as evidence and challenge for other authors [5].

References

1. Kennedy, J. (2011). Particle swarm optimization. In *Encyclopedia of machine learning* (pp. 760–766). USA: Springer.
2. Melin, P., Olivas, F., Castillo, O., Valdez, F., Soria, J., & Valdez, M. (2013). Optimal design of fuzzy classification systems using PSO with dynamic parameter adaptation through fuzzy logic. *Expert Systems with Applications, 40*(8), 3196–3206.
3. Amador-Angulo, L., & Castillo, O. (2016). Comparative study of bio-inspired algorithms applied in the design of fuzzy controller for the water tank. In *Recent developments and new*

direction in soft-computing foundations and applications (pp. 419–438). Springer International Publishing.

4. Amador-Angulo, L., Mendoza, O., Castro, J. R., Rodríguez-Díaz, A., Melin, P., & Castillo, O. (2016). Fuzzy sets in dynamic adaptation of parameters of a bee colony optimization for controlling the trajectory of an autonomous mobile robot. *Sensors, 16*(9), 1458.

5. Caraveo, C., Valdez, F., & Castillo, O. (2016). Optimization of fuzzy controller design using a new bee colony algorithm with fuzzy dynamic parameter adaptation. *Applied Soft Computing, 43,* 131–142.

6. Karaboga, D., & Basturk, B. (2007). A powerful and efficient algorithm for numerical function optimization: artificial bee colony (ABC) algorithm. *Journal of Global Optimization, 39*(3), 459–471.

7. Song, G. C., & Ryu, C. M. (2013). Two volatile organic compounds trigger plant self-defense against a bacterial pathogen and a sucking insect in cucumber under open field conditions. *International Journal of Molecular Sciences, 14*(5), 9803–9819.

8. Azar, D., Fayad, K., & Daoud, C. (2016). A combined ant colony optimization and simulated annealing algorithm to assess stability and fault-proneness of classes based on internal software quality attributes. *International Journal of Artificial Intelligence™, 14*(2), 137–156.

9. Olivas, F., Valdez, F., & Castillo, O. (2015). Dynamic parameter adaptation in Ant Colony Optimization using a fuzzy system for TSP problems. In *IFSA-EUSFLAT* (pp. 765–770).

10. Gaxiola, F., Melin, P., Valdez, F., Castro, J. R., & Castillo, O. (2016). Optimization of type-2 fuzzy weights in backpropagation learning for neural networks using GAs and PSO. *Applied Soft Computing, 38,* 860–871.

11. González, C. I., Castro, J. R., Martínez, G. E., Melin, P., & Castillo, O. (2013, June). A new approach based on generalized type-2 fuzzy logic for edge detection. *In IFSA World Congress and NAFIPS Annual Meeting (IFSA/NAFIPS), 2013 Joint* (pp. 424–429). IEEE.

12. González, C. I., Melin, P., Castro, J. R., Castillo, O., & Mendoza, O. (2016). Optimization of interval type-2 fuzzy systems for image edge detection. *Applied Soft Computing, 47,* 631–643.

13. Melin, P., Castillo, O., Gonzalez, C. I., Castro, J. R., & Mendoza, O. (2016, October). General type-2 fuzzy edge detectors applied to face recognition systems. In *Fuzzy Information Processing Society (NAFIPS), 2016 Annual Conference of the North American* (pp. 1–6). IEEE.

14. Ochoa, P., Castillo, O., & Soria, J. (2016, September). Fuzzy differential evolution method with dynamic parameter adaptation using type-2 fuzzy logic. In *2016 IEEE 8th International Conference on Intelligent Systems (IS)* (pp. 113–118). IEEE.

15. Koornneef, A., & Pieterse, C. M. (2008). Cross talk in defense signaling. *Plant Physiology, 146*(3), 839–844.

16. Laumanns, M., Rudolph, G., & Schwefel, H. P. (1998, September). A spatial predator-prey approach to multi-objective optimization: A preliminary study. In *International Conference on Parallel Problem Solving from Nature* (pp. 241–249). Berlin: Springer.

17. Law, J. H., & Regnier, F. E. (1971). Pheromones. *Annual Review of Bio-chemistry, 40*(1), 533–548.

18. Caraveo, C., Valdez, F., & Castillo, O. (2015). A new bio-inspired optimization algorithm based on the self-defense mechanisms of plants. In *Design of Intelligent Systems Based on Fuzzy Logic, Neural Networks and Nature-Inspired Optimization* (pp. 211–218). Springer International Publishing.

19. Caraveo, C., Valdez, F., & Castillo, O. (2015). Bio-inspired optimization algorithm based on the self-defense mechanism in plants. In *Advances in artificial intelligence and soft computing* (pp. 227–237). Springer International Publishing.

20. Caraveo, P. (2016, December). A new metaheuristic based on the self-defense techniques of the plants in nature. In *2016 IEEE Symposium Series on Computational Intelligence (SSCI)* (pp. 1–5). IEEE.

21. Olivas, F., Valdez, F., Castillo, O., Gonzalez, C. I., Martinez, G., & Melin, P. (2017). Ant colony optimization with dynamic parameter adaptation based on interval type-2 fuzzy logic systems. *Applied Soft Computing, 53,* 74–87.
22. Barraza, J., Melin, P., Valdez, F., & Gonzalez, C. I. (2016, July). Fuzzy FWA with dynamic adaptation of parameters. In *2016 IEEE Congress on Evolutionary Computation (CEC)* (pp. 4053–4060). IEEE.
23. Peraza, C., Valdez, F., Garcia, M., Melin, P., & Castillo, O. (2016). A new fuzzy harmony search algorithm using fuzzy logic for dynamic parameter adaptation. *Algorithms, 9*(4), 69.
24. Pérez, J., Valdez, F., & Castillo, O. (2017). Modification of the bat algorithm using type-2 fuzzy logic for dynamical parameter adaptation. In *Nature-inspired design of hybrid intelligent systems* (pp. 343–355). Springer International Publishing.
25. Perez, J., Valdez, F., Castillo, O., & Roeva, O. (2016, September). Bat algorithm with parameter adaptation using interval type-2 fuzzy logic for benchmark mathematical functions. In *2016 IEEE 8th International Conference on Intelligent Systems (IS)* (pp. 120–127). IEEE.
26. Perez, J., Valdez, F., Castillo, O., Melin, P., Gonzalez, C., & Martinez, G. (2017). Interval type-2 fuzzy logic for dynamic parameter adaptation in the bat algorithm. *Soft Computing, 21* (3), 667–685.
27. Teodorovic, D., Bee colony optimization (BCO). (2009). In C. P. Lim, L. C. Jain, & S. Dehuri (Eds.), *Innovations in swarm intelligence* (pp. 39–60). Berlin: Springer. (65, 215).
28. Harmanani, H. M., Drouby, F., & Ghosn, S. B. (2009, March). A parallel genetic algorithm for the open-shop scheduling problem using deterministic and random moves. In *Proceedings of the 2009 Spring Simulation Multiconference* (p. 30). Society for Computer Simulation International.

Chapter 2
Theory and Background

In the literature there are some published works where the authors use the predatory prey mathematical model, to model problems, but the main difference of our proposal against the existing works is that we propose an optimization algorithm, which is iterative and applying evolution processes to improve the adaptation to the habitat that belongs.

In some recent published works previous authors show that the animals as well as plants defend themselves against invading pathogenic microorganisms utilizing cationic antimicrobial peptides, which rapidly kill various microbes without exerting toxicity against the host. Physicochemical peptide–lipid interactions provide attractive mechanisms for innate immunity. Many of these peptides form cationic amphipathic secondary structures, typically α-helices and β-sheets, which can selectively interact with anionic bacterial membranes by the aid of electrostatic interactions.

In [1, 2] the authors applied the traditional predator prey model to approximate the solution of nonlinear functions, also in [3, 4] the predatory prey method is used as a prediction model, which can be used in ecology if the evolution of the species were in shorter time cycles, and in [5–7] ant colony optimization with dynamic parameter based on interval Type-2 fuzzy logic systems are presented. In the case of designing type-2 fuzzy controllers for particular applications, the use of bio-inspired optimization methods have helped in the complex task of finding the appropriate parameter values and structure of the fuzzy systems. In this review, we consider the application of genetic algorithms, particle swarm optimization and ant colony optimization as three different paradigms that help in the design of optimal type-2 fuzzy controllers. Also in [7], Ant Colony Optimization using a fuzzy system for TSP problems with dynamic parameter adaptation was proposed. The author performs a modification of the ant colony algorithm, using FLS to dynamically adapt alpha and rho values, applied to the Travelling Salesman Problem (TSP). Other algorithms have also been combined with FLS for example, in [5] the differential evolution algorithm is modified with FL for the adjustment of values and applied to mathematical functions benchmark. Also in [8] an algorithm based on the harmony

© The Author(s), under exclusive license to Springer Nature Switzerland AG 2019
C. Caraveo et al., *A New Bio-inspired Optimization Algorithm Based on the Self-defense Mechanism of Plants in Nature*, SpringerBriefs in Computational Intelligence, https://doi.org/10.1007/978-3-030-05551-6_2

search was optimized with fuzzy logic and the performance of the algorithm with fuzzy approach was tested in benchmark mathematical functions.

A new variant of the bio-inspired algorithm based on bats is presented in [9–11], namely an Interval Type-2 fuzzy logic for dynamic parameter adaptation in the bat algorithm, the author of this work uses interval Type-2 fuzzy logic to adapt some important parameters in the algorithm and is used to optimize a set of benchmark mathematical functions. In [12, 13] the authors state that: the plants are equipped with an array of defense mechanisms to protect themselves against attack by herbivorous insects and microbial pathogens. Some of these defense mechanisms are preexisting, whereas others are only activated upon insect or pathogen invasion.

Induced defense responses entail fitness costs. Therefore, plants possess elaborate regulatory mechanisms that efficiently coordinate the activation of attacker specific defenses so that fitness costs are minimized, while optimal resistance is attained. In [14–16] the authors of this work claim that some of the chemical reactions that release plants can be used to combat pest problems in agriculture. The previous work above mentioned is the most similar to the work proposed in this book, but not the same.

There are also other published works where the predator prey model is used to solve different optimization problems, for example in [17] Predator–Prey Brain Storm Optimization (PPBSO) for DC Brushless Motor, the Predator–prey concept is adopted to better utilize the global information and improve the swarm diversity during the evolution process. The proposed algorithm is applied to solve the optimization problems in an electromagnetic field. The comparative results demonstrate that both PPBSO and BSO can succeed in optimizing design variables for a DC brushless motor to maximize its efficiency. And they also propose an innovative algorithm called Brain Storm Optimization (BSO) is a newly-developed swarm intelligence optimization algorithm inspired by a human being's behavior of brainstorming.

References

1. Duffy, B., Schouten, A., & Raaijmakers, J. M. (2003). Pathogen self-defense: Mechanisms to counteract microbial antagonism. *Annual Review of Phytopathology, 41*(1), 501–538.
2. García-Garrido, J. M., & Ocampo, J. A. (2002). Regulation of the plant defense response in arbuscular mycorrhizal symbiosis. *Journal of Experimental Botany, 53*(373), 1377–1386.
3. Bennett, R. N., & Wallsgrove, R. M. (1994). Secondary metabolites in plant defense mechanisms. *New Phytologist, 127*(4), 617–633.
4. Berryman, A. A. (1992). The origins and evolution of predator-prey theory. *Ecology, 73*(5), 1530–1535.
5. Neyoy, H., Castillo, O., & Soria, J. (2013). Dynamic fuzzy logic parameter tuning for ACO and its application in TSP problems. *Recent advances on hybrid intelligent systems* (pp. 259–271). Berlin, Heidelberg: Springer.
6. Olivas, F., Valdez, F., & Castillo, O. (2015). Dynamic parameter adaptation in ant colony optimization using a fuzzy system for TSP problems. In *IFSA-EUSFLAT* (pp. 765–770).

7. Olivas, F., Valdez, F., Castillo, O., Gonzalez, C. I., Martinez, G., & Melin, P. (2017). Ant colony optimization with dynamic parameter adaptation based on interval type-2 fuzzy logic systems. *Applied Soft Computing, 53*, 74–87.
8. Peraza, C., Valdez, F., Garcia, M., Melin, P., & Castillo, O. (2016). A new fuzzy harmony search algorithm using fuzzy logic for dynamic parameter adaptation. *Algorithms, 9*(4), 69.
9. Pérez, J., Valdez, F., & Castillo, O. (2017). Modification of the bat algorithm using type-2 fuzzy logic for dynamical parameter adaptation. *Nature-inspired design of hybrid intelligent systems* (pp. 343–355). Cham: Springer International Publishing.
10. Perez, J., Valdez, F., Castillo, O., & Roeva, O. (2016, September). Bat algorithm with parameter adaptation using interval type-2 fuzzy logic for benchmark mathematical functions. In *2016 IEEE 8th International Conference on Intelligent Systems (IS)* (pp. 120–127). IEEE.
11. Perez, J., Valdez, F., Castillo, O., Melin, P., Gonzalez, C., & Martinez, G. (2017). Interval type-2 fuzzy logic for dynamic parameter adaptation in the bat algorithm. *Soft Computing, 21*(3), 667–685.
12. Vivanco, J. M., Cosio, E., Loyola-Vargas, V. M., & Flores, H. E. (2005). Mecanismos químicos de defensa en las plantas. *Investigación y ciencia, 341*(2), 68–75.
13. Wang, M. B., & Metzlaff, M. (2005). RNA silencing and antiviral defense in plants. *Current Opinion in Plant Biology, 8*(2), 216–222.
14. Pieterse, C. M., & Dicke, M. (2007). Plant interactions with microbes and insects: From molecular mechanisms to ecology. *Trends in Plant Science, 12*(12), 564–569.
15. Song, G. C., & Ryu, C. M. (2013). Two volatile organic compounds trigger plant self-defense against a bacterial pathogen and a sucking insect in cucumber under open field conditions. *International Journal of Molecular Sciences, 14*(5), 9803–9819.
16. Tollsten, L., & Muller, P. M. (1996). Volatile organic compounds emitted from beech leaves. *Phytochemistry, 43*, 759–762.
17. Molina, D., & Herrera, F. (2015, May). Iterative hybridization of DE with local search for the CEC'2015 special session on large scale global optimization. In *2015 IEEE congress on evolutionary computation (CEC)* (pp. 1974–1978). IEEE.

Chapter 3
Self-defense of the Plants

Defense mechanisms (or coping strategies) are automatic natural processes that protect individuals against external or internal threats in general. In nature, plants are exposed to many invading predators, such as insects, fungi, bacteria and viruses [1–4]. Plants do not have mobility, therefore their survival depends entirely on their immune system and other strategies or evolutionary adaptation strategies developed to prevent death or extinction of the plants [5–7]. This suggests that the defense mechanisms of the plants are very effective to lock or counteract an infection and keep away predators. Additionally, it has been shown that plants are able to react to different stimuli, such as light intensity, quantity and quality of water or the presence of some toxic substances around. Plants have a nonlinear behavior pattern, which acts directly to any external stimulus.

When the plant suffers from aggression, it triggers a series of chemical reactions that are liberated into the air, these substances released, attract the predator's natural enemies which are attacking the plant. In Fig. 3.1 a general diagram of the plant defense process when it detects the presence of an invading organism is presented.

In Fig. 3.1 a general diagram of the process of self-defense of the plant is presented, when it detects the attack by a predator for example: insects, bacteria, fungi [6, 8]. In this case, a plant releases a series of chemical reactions and the products are released into the air, this attracts different types of insects, such as pollinating insects to achieve the reproduction before death, and preserve their species against extinction. These can also be insects like seed dispensers, or the natural enemy of the predator is attacking the plant. In nature, the plants have different methods of biological reproduction for example, Pollination, graft and cloning, these are some of the most common methods of reproduction, and the methods are described in more detail below.

Graft: A method of vegetative reproduction of plants, where a portion of tissue extracted from a plant is inserted into another, in order that both grow as a single

C. Caraveo et al., *A New Bio-inspired Optimization Algorithm Based on the Self-defense Mechanism of Plants in Nature*, SpringerBriefs in Computational Intelligence, https://doi.org/10.1007/978-3-030-05551-6_3

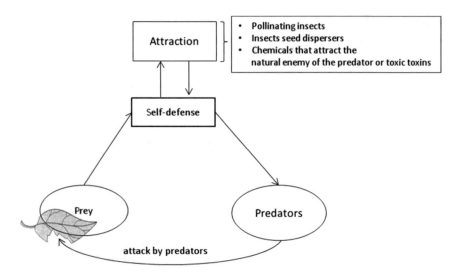

Fig. 3.1 General scheme of the predator attack on plants

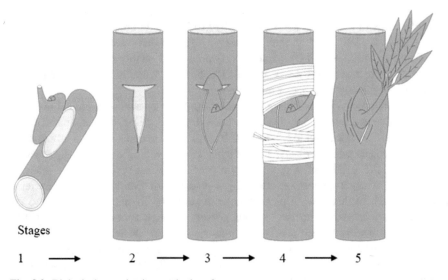

Fig. 3.2 Biological reproduction method graft

organism and share their features. The graft method of reproduction is possible only between species of plants belonging to the same species, so that their tissues can be compatible and ensure the survival of the species, and this method is illustrated in Fig. 3.2.

In Fig. 3.2 we find a graphical representation of the graft method of vegetative reproduction, where a fragment is taken from one plant and inserted into another plant and that automatically starts a process of merging between the two tissues to grow as a single plant and inheriting their different characteristics.

Cloning: In a biological sense cloning is meant by obtaining genetically identical individuals. Also in nature there are clones, from the descendants of those organisms that reproduce asexually, plants can be propagated from a fragment of the plant. This method of biological reproduction allows the best plants or individuals to reproduce and preserve their characteristics of which are inherited to other generations. Figure 3.3 illustrates the cloning process.

In Fig. 3.3 an illustration of the reproductive cloning method is presented, and this method allows providing to the next generation the best genes and characteristics of plants that can be preserved throughout time.

Pollination: Pollination is a biological reproduction method used by plants to send grains of pollen from one plant (flower) to another plant (flower). In order for this process to be performed it depends on several factors, in this case the most common are:

Pollination by insects (biotic): This process of reproduction is totally dependent on birds and insect pollinators, in fact pollination is more common using bees, when a bee visits a plant to collect honey, it also collects pollen and this pollen is transported to the following plants the bee visits on its way in search for food. This process is also performed by other insects such as butterflies, bats, ants and other animals [9, 10]. In Fig. 3.4 we can find an illustration of the process of pollination by pollinating insects, where an insect randomly decides to visit neighboring plant.

Pollination by air (abiotic): In this case, the pollen produced by plants is transported to other locations using air currents; in this case the air is totally responsible for carrying the pollen from one flower to another flower [9, 11].

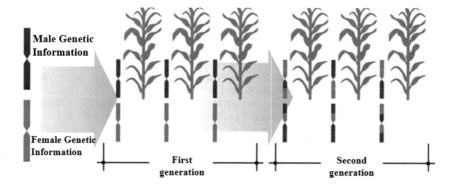

Fig. 3.3 Biological reproduction cloning method

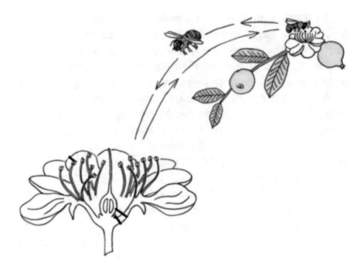

Fig. 3.4 Reproduction method by pollination

References

1. Pieterse, C. M., & Dicke, M. (2007). Plant interactions with microbes and insects: From molecular mechanisms to ecology. *Trends in Plant Science, 12*(12), 564–569.
2. Ryan, C. A., & Jagendorf, A. (1995). Self-defense by plants. *Proceedings of the National Academy of Sciences, 92*(10), 4075.
3. Vivanco, J. M., Cosio, E., Loyola-Vargas, V. M., & Flores, H. E. (2005). Mecanismos químicos de defensa en las plantas. *Investigación y ciencia, 341*(2), 68–75.
4. Yoshida, T., Jones, L. E., Ellner, S. P., Fussmann, G. F., & Hairston, N. G. (2003). Rapid evolution drives ecological dynamics in a predator–prey system. *Nature, 424*(6946), 303–306.
5. Cruz, J. M. L., & González, G. B. (2008). Modelo depredador-presa. *Revista de Ciencias Básicas UJAT, 7*(2), 25–34.
6. Paré, P. W., & Tumlinson, J. H. (1999). Plant volatiles as a defense against insect herbivores. *Plant Physiology, 121*(2), 325–332.
7. Peraza, C., Valdez, F., Garcia, M., Melin, P., & Castillo, O. (2016). A new fuzzy harmony search algorithm using fuzzy logic for dynamic parameter adaptation. *Algorithms, 9*(4), 69.
8. Ordeñana, K. M. (2002). Mecanismos de defensa en las interacciones planta-patógeno. Revista Manejo Integrado de Plagas. *Costa Rica, 63,* 22–32.
9. Yang, X. S. (2012). Flower pollination algorithm for global optimization. In *Unconventional computation and natural computation* (pp. 240–249). Berlin, Heidelberg: Springer.
10. Yang, X. S., Karamanoglu, M., & He, X. (2014). Flower pollination algorithm: A novel approach for multi-objective optimization. *Engineering Optimization, 46*(9), 1222–1237.
11. Dafni, A., Kevan, P. G., & Husband, B. C. (2005). *Practical pollination biology.*

Chapter 4
Predator-Prey Model

The organisms live in communities, forming intricate relationships of interaction, where each species directly or indirectly depend on the presence of the other. One of the tasks of Ecology is to develop a theory of community organization for understanding the causes of diversity and mechanisms of interaction. In this book, we consider the interaction of two species whose populations size at time t are $x(t)$ and $y(t)$ [1–5]. Furthermore, we assume that the change in population size can be written as:

$$\frac{dy}{dt} = I(x, y) \tag{4.1}$$

$$\frac{dx}{dt} = P(x, y) \tag{4.2}$$

There are different kinds of biological interaction that can be represented mathematically with the system of Eq. 4.3. As $P(x, y)$ and $I(x, y)$ are determining the growth rate of each of the populations; there is the case where one of these species is fed from the other, then the system of survival is given by: Eq. 4.3.

$$\begin{aligned} P_y(x, y) &< 0 \\ I_x(x, y) &> 0 \end{aligned} \tag{4.3}$$

That is, the change of the prey population relative to the predator decreases and the change of the predator population relative to the prey increases. These are some of the conditions that must meet a set of predator prey equations, like Eq. 4.3.

C. Caraveo et al., *A New Bio-inspired Optimization Algorithm Based on the Self-defense Mechanism of Plants in Nature*, SpringerBriefs in Computational Intelligence, https://doi.org/10.1007/978-3-030-05551-6_4

4.1   Analysis of the Lotka and Volterra Model

This model is based on the following assumptions.

The population grows proportionally to its size, and has enough space and food. If this happens and x(t) represents the prey population (in the absence of predators), then the population growth is given by:

$$\frac{dx}{dt} = ax, \quad a > 0,$$
$$x(t) = x_0 e^{at}. \tag{4.4}$$

The population of prey in the absence of the predator grows exponentially.

The predator y(t) only feeds on the prey x(t). Thus, if there is no prey, their size decreases with a rate proportional to its population is represented by Eq. 4.5

$$\frac{dy}{dt} = -dy, \quad d > 0,$$
$$y(t) = y_0 e^{-dt} \tag{4.5}$$

The population of predators in the absence of prey decreases exponentially to extinction.

The number of encounters between predator and prey is proportional to the product of their populations. Each of the number of encounters favor predators and reduces the number of prey.

The presence of prey helps the growth of the predator and is represented by Eq. (4.6)

$$Gp = c \times y \quad c > 0. \tag{4.6}$$

While the interaction between them, reduces the growth of prey is represented by Eq. (4.7)

$$GPy = -b \times y \quad b > 0. \tag{4.7}$$

Under the above hypothesis, we have a model of interaction between x(t) and y(t) is given by the following system: Eqs. (4.8) and (4.9)

$$\frac{dx}{dt} = \alpha x - \beta xy \tag{4.8}$$

$$\frac{dy}{dt} = -\delta xy + \lambda y \tag{4.9}$$

where:

x is the number of prey.

y is the number of predators.

$\frac{dx}{dt}$ is the growth of the population of prey time t.

$\frac{dy}{dt}$ is the growth of the population of predator at time t.

α It represents the birth rate of prey in the absence of predator.

β It represents the death rate of predators in the absence of prey.

δ Measures the susceptibility of prey.

λ Measures the ability of predation.

References

1. Caraveo, C., Valdez, F., & Castillo, O. (2015). Bio-inspired optimization algorithm based on the self-defense mechanism in plants. In *Advances in artificial intelligence and soft computing* (pp. 227–237). Springer International Publishing.
2. Caraveo, C., Valdez, F., Castillo, O., & Melin, P. (2016, December). A new metaheuristic based on the self-defense techniques of the plants in nature. In *2016 IEEE Symposium Series on Computational Intelligence (SSCI)* (pp. 1–5). IEEE.
3. Lez-Parra, G. G., Arenas, A. J., & Cogollo, M. R. (2013). Numerical-analytical solutions of predator-prey models. *WSEAS Transactions on Biology & Biomedicine, 10*(2), 1–7.
4. Silva, A., Neves, A., & Costa, E. (2002, September). An empirical comparison of particle swarm and predator prey optimisation. In *Irish Conference on Artificial Intelligence and Cognitive Science* (pp. 103–110). Berlin, Heidelberg: Springer.
5. Xiao, Y., & Chen, L. (2001). Modeling and analysis of a predator–prey model with disease in the prey. *Mathematical Biosciences, 171*(1), 59–82.

Chapter 5
Proposed Method

In this book we propose a new optimization method, which is bio-inspired on the self-defense process of plants in nature. This new algorithm is created in order to solve complex optimization problems with a minimal computer use and reducing runtime of the algorithm. For the development of the proposed algorithm, the predator prey model proposed by Lotka-Volterra was used as the main basis. As explained in Chap. 2, plants are able to react to different stimuli, such as: Air, sun, water, darkness, and threats by different predators, such as providing shelter for other animals to protect them from different predators that feed on them [1–3].

In the following Figure, we find in more details on the procedure that the algorithm performs internally in the proposed approach, see Fig. 5.1.

The description of the stages and operations of the algorithm is presented on Table 5.1

In Fig. 5.2, a general diagram of the algorithm and operations are performed, also the biological representation of the reproduction used in this proposal is presented.

In Fig. 5.2, an illustration is presented, where we can observe the traditional model of predator prey and the approach proposed by the authors in this work, and as we can notice we are focusing on the population of prey. In this case, the plants that are subjected to a process of evolution to develop the techniques of confrontation, for the process of evolution is necessary to apply some biological operators as shown in Fig. 5.3.

5.1 Biological Reproduction Method and Proposed Approach

This section describes in more detail the internal representation of the reproduction methods, in the proposed algorithm.

© The Author(s), under exclusive license to Springer Nature Switzerland AG 2019
C. Caraveo et al., *A New Bio-inspired Optimization Algorithm Based on the Self-defense Mechanism of Plants in Nature*, SpringerBriefs in Computational Intelligence, https://doi.org/10.1007/978-3-030-05551-6_5

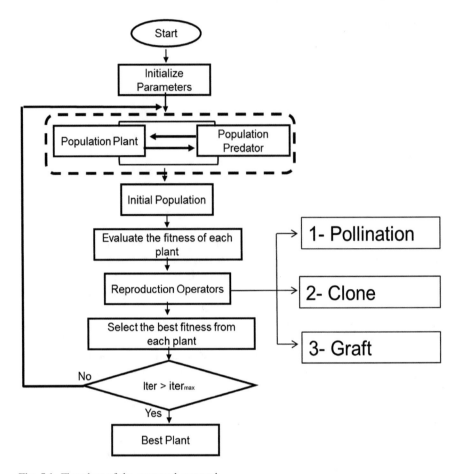

Fig. 5.1 Flowchart of the proposed approach

Reproduction method by the Graft process: in this case a stem of a plant is used and is encrusted into another to generate an alteration in the structure of the plant.

In the previous Section, the method of reproduction by graft was described, this method is one of the most used, and it allows us to preserve the best characteristics of an individual and inherit the future generations of new plants. In Fig. 5.3 a general diagram of the process performed in nature and the internal process of the algorithm (our proposal) is presented. The plant with higher fitness value of the population is obtained, and then a second plant is taken from the population, in this case one of the plants with a low fitness value. Both plants are combined to improve the characteristics of the plant with lower fitness value, with a probability $p(x)\ \varepsilon\ [0, 1]$, a local or global graft is determined.

A global graft consists in inserting some of the best characteristics of the plant in all the subpopulation and a local graft consists in inserting characteristics of the best plant only in another plant, it is also called a local search and global search, all this

Table 5.1 Description of the stages of the proposed meta-heuristic

Step	Description
Start	The algorithm begins
Initialize parameters	The parameters of α, β, λ, δ, must be initialized before starting the algorithm, and these parameters help control the growth of both populations in the absence of prey or predators. Also other factors, such as the amount of food that can be consumed and the number of confrontations between the two populations, also defined the reproduction method to use, such as: graft, pollinated and clone
Plants and predator	In this step these two populations interact with each other and the initial population of plants is generated
Initial populations	After each encounter between prey and predator, a plant population is generated
Evaluate the fitness of each plant	In this step a pre-evaluation is performed to detect the fitness value of each plant, and the characteristics of the plant can be inherited to another plant, according to the reproduction method used
Reproduction operators	In this phase of the algorithm, the biological reproduction operator is applied, and in this case we only consider three: reproduction by cloning, graft and pollination, when the algorithm started the user manually selects the reproduction method to use
Select the best fitness from each plant	In this step of the algorithm all plants are evaluated using a method of selection, in this case we use the roulette selection type, and of this population the best plants are selected to replace the worst plants in the new populations generated by the encounters between the two species
Iter > itermax	In this step a condition is verified to validate if the maximum number of iterations is complete, otherwise returns to step number 4 and continue with the iterations
End	End the search process of the algorithm

with the purpose of maintaining a better balance between exploitation and exploration in the proposed algorithm. In the following paragraph the method of biological reproduction by pollination is described.

Reproduction method by the Pollination process: The transport of pollen is performed by air or by animals; the plants produce millions of grains of pollen that are transported to other plants in the air. In the case of animals, the plants attract insects and birds using flower colors, producing nectar, or produce volatile pollen that is transported by air. In Fig. 5.4 a representation of the natural process and the proposal is presented.

In Fig. 5.4 a general diagram is presented and the process of the proposed method, the plant with greater fitness value, is selected to pollinate other plants as shown, then a second plant is taken from the population.

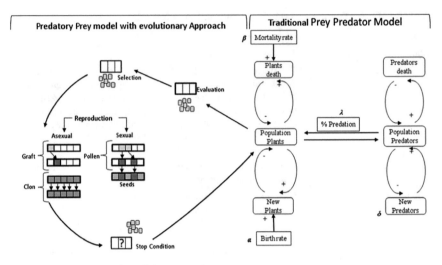

Fig. 5.2 General illustration of the proposal

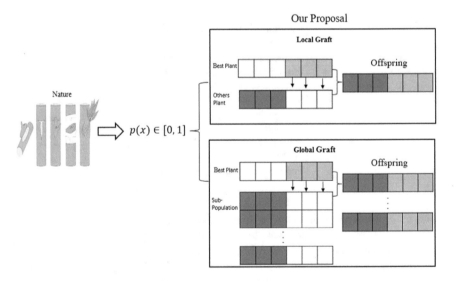

Fig. 5.3 Graft reproduction method

In this case the plant with a higher fitness value is used to pollinate neighboring plants, with a probability $p(x) \; \varepsilon \; [0, 1]$, is determined if the pollinating insect visit plants that have lower or higher distance from its current value, for this reproduction method we are using as a basis the Levy flights [4, 5] as shown in Fig. 5.5.

Lévy flights, named in honor of French mathematician Paul Pierre Lévy, are a type of random walk where the length of the jumps is distributed according to a probability of distribution [4]. In this case for each pollinator insect a Levy flight is

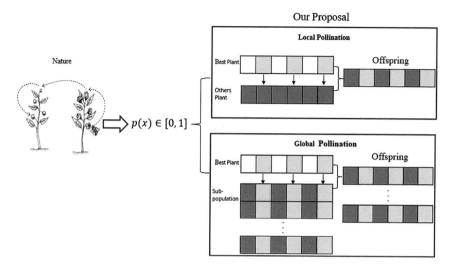

Fig. 5.4 Pollination reproduction method

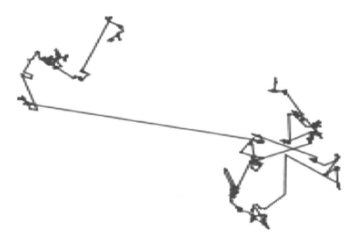

Fig. 5.5 Levy flight illustration

assigned and the length of the pollinating insect flight is determined with a certain probability, this in order to maintain a better balance between exploitation and exploration in the proposed algorithm.

Reproduction by the cloning method process: The Cloning method is a method used to preserve the total characteristics of an individual, in this case, this method is used to preserve the plant with greater fitness value of the population and inherit all the characteristics in the new offspring. In Fig. 5.6 a general diagram of the original approach and the proposed approach is presented.

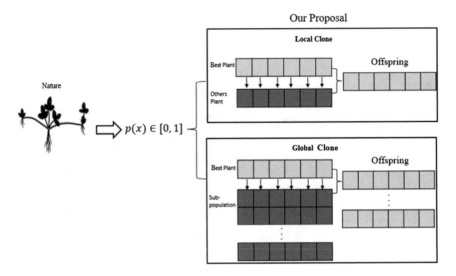

Fig. 5.6 Reproduction method by Cloning

In Fig. 5.6 we can find a general scheme of the natural process and internal process of the proposed algorithm. In this case, during the iterations of the algorithm the fitness value of each plant is measured, the plant with a low fitness value is cloned with the characteristics of the plant with greater fitness value, with a probability $p(x)\ \varepsilon\ [0, 1]$, and it is also determined whether to apply a local or global cloning.

References

1. Caraveo, C., Valdez, F., & Castillo, O. (2015). Bio-inspired optimization algorithm based on the self-defense mechanism in plants. In *Advances in artificial intelligence and soft computing* (pp. 227–237). Springer International Publishing.
2. Koornneef, A., & Pieterse, C. M. (2008). Cross talk in defense signaling. *Plant Physiology, 146* (3), 839–844.
3. Laumanns, M., Rudolph, G., & Schwefel, H. P. (1998, September). A spatial predator-prey approach to multi-objective optimization: A preliminary study. In *International Conference on Parallel Problem Solving from Nature* (pp. 241–249). Berlin, Heidelberg: Springer.
4. Yang, X. S. (2012). *Flower pollination algorithm for global optimization. In Unconventional computation and natural computation* (pp. 240–249). Berlin, Heidelberg: Springer.
5. Yang, X. S., & Deb, S. (2009, December). Cuckoo search via Lévy flights. In *2009 World Congress on Nature & Biologically Inspired Computing, NaBIC 2009* (pp. 210–214). IEEE.

Chapter 6
Case Studies

In this chapter we present the different cases of study that are used and analyzed to test the performance and efficiency of the meta-heuristics developed in this book. Five case studies were tested using the optimization algorithm bio-inspired on the self-defense mechanisms of the plants in nature, and the case studies are presented below.

6.1 Case Study One: Optimization of Benchmark Mathematical Functions

The performance of the proposed meta-heuristic of optimization is tested on the following set of benchmark mathematical functions [1, 2]. The name and the mathematical definition of the functions used in this book are shown below:

Powell Function

The function is usually evaluated on the hypercube $x_i \in [-4, 5]$, for all i = 1, ..., d.

$$f(\mathbf{x}) = \sum_{i=1}^{d/4} \left[(x_{4i-3} + 10x_{4i-2})^2 + 5(x_{4i-1} - x_{4i})^2 \cdots \right. \\ \left. + (x_{4i-2} - 2x_{4i-1})^4 + 10(x_{4i-3} - x_{4i})^2 \right] \tag{6.1}$$

Ackley Function

The function is usually evaluated on the hypercube $x_i \in [-32.768, 32.768]$, for all i = 1, ..., d, although it may also be restricted to a smaller domain.

© The Author(s), under exclusive license to Springer Nature Switzerland AG 2019
C. Caraveo et al., *A New Bio-inspired Optimization Algorithm Based
on the Self-defense Mechanism of Plants in Nature*, SpringerBriefs
in Computational Intelligence, https://doi.org/10.1007/978-3-030-05551-6_6

$$f(\mathbf{x}) = -a \cdot \exp\left(-b \cdot \sqrt{\frac{1}{n}\sum_{i=1}^{n}(x_i^2)} - \exp\left(\frac{1}{n}\sum_{i=1}^{n}\cos(cx_i)\right)\right) + a + \exp(1) \quad (6.2)$$

Griewank Function

The function is usually evaluated on the hypercube $x_i \in [-600, 600]$, for all $i = 1, ..., d$.

$$f(\mathbf{x}) = \sum_{i=1}^{d}\frac{x_i^2}{4000} - \prod_{i=1}^{d}\cos\left(\frac{x_i}{\sqrt{i}}\right) + 1 \quad (6.3)$$

Rastrigin Function

The function is usually evaluated on the hypercube $x_i \in [-5.12, 5.12]$, for all $i = 1, ..., d$.

$$f(\mathbf{x}) = \sum_{i=1}^{d}\frac{x_i^2}{4000} - \prod_{i=1}^{d}\cos\left(\frac{x_i}{\sqrt{i}}\right) + 1 \quad (6.4)$$

Sphere Function

The function is usually evaluated on the hypercube $x_i \in [-5.12, 5.12]$, for all $i = 1, ..., d$.

$$f(x) = \sum_{i=1}^{n}x_i^2 \quad (6.5)$$

Levy Function

The function is usually evaluated on the hypercube $x_i \in [-10, 10]$, for all $i = 1, ..., d$.

$$f(x) = \sin^2(\pi\omega_1) + \cdots \sum_{i=1}^{d-1} (\omega_1 - 1)^2[1 + \sin^2(\pi\omega_i + 1)]$$

$$+ (\omega_d - 1)^2[1 + \sin^2(2\pi\omega_d)]$$

(6.6)

Axis Parallel Hyper-Ellipsoid function

The function can be defined on any input domain but it is usually evaluated on $x_i \in [-10, 10]$ for $i = 1, \ldots, n$.

$$f(x) = \sum_{i=1}^{n} ix_i^2$$

(6.7)

Rotated Hyper-Ellipsoid Function

The function is usually evaluated on the hypercube $x_i \in [-65.536, 65.536]$, for all $i = 1, \ldots, d$.

$$f(x) = \sum_{i=1}^{d} \left(\sum_{j=1}^{i} x_j \right)^2$$

(6.8)

Dixon-Price Function

The function is usually evaluated on the hypercube $x_i \in [-10, 10]$, for all $i = 1, \ldots, d$.

$$f(x) = (x_1 - 1)^2 + \sum_{i=2}^{d} i(2x_i^2 - x_{i-1})^2$$

(6.9)

Zakharov Function

The function is usually evaluated on the hypercube $x_i \in [-5, 10]$, for all $i = 1, \ldots, d$.

$$f(x) = \sum_{i=1}^{d} x_i^2 + \left(\sum_{i=1}^{d} 0.5ix_i \right)^2 + \left(\sum_{i=1}^{d} 0.5ix_i \right)^4$$

(6.10)

6.1.1 Simulations Results

The optimization algorithm bio-inspired on the self-defense mechanisms of plants, was used to optimize benchmark mathematical functions, for the case study described on the previous section. It was tested for a set of 10 mathematical functions, for 30, 50 and 100 variables, where the objective value is approximate to zero. The initial sizes of both populations (plants, predators) are defined by the user, the parameters (α, β, λ, δ) are also defined by the user, and for the model of Lotka and Volterra the following parameter values are recommended $\alpha = 0.4$, $\beta = 0.37$, $\lambda = 0.3$, $\delta = 0.05$.

For this problem the values for the variables are manually moved in the following ranges α, β, λ, δ, they are in [0, 1] and using the three methods of biological reproduction proposed in this book with the purpose of observing the behavior of the algorithm and determine what values and what ranges are optimal for the proposed algorithm, the results obtained are shown in Table 6.1.

In Table 6.1 we can find the results after applying the algorithm to the proposed mathematical functions. We decided to show the most significant data for the 30 experiments performed using different methods of reproduction and different number of variables. The most significant data are: Functions, Reproduction Method, Variables, Best, Worst, σ, Average. We can notice that the proposed approach demonstrates good performance in some of the proposed functions in this case study, such as: Dixon, Rosenbrock, and Levy. In these functions the algorithm performance was low for some numbers of variables, but it is observed that when the number of variables is high the algorithm has difficulty in approximating the value of the function to zero. In some experiments the algorithm achieved very good results but not in others, this behavior of the algorithm causes a standard deviation and average values with very high results.

The proposed algorithm is under improvements and adaptations in order to compete against existing algorithms in the literature; with the experiments we find optimal ranges values for the algorithm for this problem. The ranges of the parameters are the following, $\alpha \in [0.3, 0.7]$, $\beta \in [0.1, 0.4]$, $\lambda \in [0.2, 0.3]$, $\delta \in [0.01, 0.05]$, and for these ranges of values found for the parameters, the algorithm offers us greater stability and balance in the exploration of solutions for this case study.

6.1.2 Statistical Comparison

In conclusion, in this book we needed to perform the statistical comparison between the performance of different methods of biological reproduction used in the proposed algorithm. In this statistical comparison, we consider only two methods in the comparison which are the more efficient according to the criteria of the experts. The statistical test used for the comparison is the z test, whose parameters are defined in Table 6.2. With the parameters of Table 6.2, we applied the statistical test for the

Table 6.1 Experimental results for 30, 50 and 100 variables

Functions	Rep. method	Dim	Best	Worst	σ	Average
Sphere	Pollination	30	1.77E−36	6.73E−15	1.23E−15	2.93E−16
		50	3.15E−76	5.92E−13	1.08E−13	1.98E−14
		100	2.49E−89	1.05E−24	2.21E−25	5.79E−26
	Clone	30	2.16E−43	6.415180	1.832414	0.709209
		50	2.1E−162	12.13978	4.729409	3.009083
		100	4.7E−205	42.73954	9.050110	2.728051
	Graft	30	4.6E−277	7.51464	1.473169	0.524378
		50	2.21E−67	0.92936	0.287375	0.136370
		100	1.25E−71	0.83502	0.245027	0.115840
Ackley	Pollination	30	8.88E−16	7.54E−14	1.67E−14	5.27E−14
		50	8.88E−16	1.21E−13	2.03E−14	9.14E−14
		100	4.89E−12	0.096420	0.018775	0.004928
	Clone	30	8.88E−16	0.881812	0.237640	0.154707
		50	8.88E−16	0.679788	0.211417	0.133736
		100	8.88E−16	3.487716	0.877724	0.325331
	Graft	30	3.19E−90	32.09056	7.669544	0.138961
		50	3.31E−16	2.915703	1.147050	0.896195
		100	6.51E−54	724.7575	1.247546	0.943876
Rastrigin	Pollination	30	2.27E−20	0.000488	8.92E−05	1.62E−05
		50	0	0.170530	0.031133	0.005688
		100	0	1.13E−12	3.928−13	4.88E−13
	Clone	30	0	28.02403	5.255146	1.750358
		50	0	14.03068	3.252023	1.573388
		100	0	13.84076	3.887058	2.715852
	Graft	30	0	0.723871	0.229009	0.155026
		50	0	0.820359	0.211544	0.110639
		100	0	0.957874	0.344052	0.388034
Griewank	Pollination	30	0	0.010177	0.001858	0.000339
		50	0	0.0155	0.002831	0.000517
		100	0	0.008011	0.001462	0.000267
	Clone	30	0	0.251355	0.045857	0.011854
		50	0	0.331116	0.084452	0.036599
		100	0	0.395976	0.092751	0.029499
	Graft	30	0	1.028922	0.198676	0.073713
		50	0	0.878538	0.166315	0.052049
		100	0	1.090982	0.211168	0.091268

(continued)

Table 6.1 (continued)

Functions	Rep. method	Dim	Best	Worst	σ	Average
Powell	Pollination	30	3.88E−48	0.001196	0.000219	5.34E−05
		50	7.34E−33	0.002363	0.000454	0.000138
		100	1.07E−26	0.012232	0.002971	0.001125
	Clone	30	2.89E−54	17.62272	4.464192	1.961976
		50	4.43E−97	14.55935	3.523362	1.320921
		100	2.21E−72	2.466369	0.634546	0.395370
	Graft	30	1.12E−93	0.967712	0.243042	0.120675
		50	3.37E−67	6.751616	1.428210	0.530652
		100	5.9E−113	0.938540	0.378009	0.282427
Rosenbrock	Pollination	30	1.9187	28.71735	8.618153	19.58295
		50	13.6763	111.4038	13.81014	44.69458
		100	14.5684	116.545	14.25647	45.6587
	Clone	30	1.91872	28.71735	8.61815	19.58295
		50	13.6763	111.4038	13.81014	44.69458
		100	14.3776	111.4661	14.46581	48.6644
	Graft	30	1.91872	28.7173	8.618153	19.58295
		50	0.79570	44.93570	15.54095	35.36125
		100	2.0711	29.836	9.23035	17.74190
Sum Square	Pollination	30	3.9E−130	1.5E−100	3.3E−101	1.0E−101
		50	4.3E−109	1.14E−73	2.55E−74	8.41E−75
		100	3.28E−93	1.53E−47	2.83E−48	6.68E−49
	Clone	30	0.017	0.0774	0.018171	0.047073
		50	0.0248	0.9359	0.280534	0.4549
		100	0.1498	3.8847	1.037645	2.55095
	Graft	30	3.19E−90	8.12853	1.561559	0.483303
		50	3.31E−16	28.3503	5.295961	1.706496
		100	6.51E−54	6.12060	1.314545	0.574652
Zakharov	Pollination	30	9.23E−50	3.90E−20	7.11E−21	1.48E−21
		50	2.33E−48	1.62E−10	3.90E−11	1.43E−11
		100	2.58E−77	0.103878	0.02099	0.00545
	Clone	30	0.0039	0.9575	0.29827	0.46026
		50	0.014	0.9135	0.25343	0.48358
		100	0.0918	2.9355	0.75097	1.39391
	Graft	30	3.13E−16	0.013715	0.00297	0.00076
		50	1.73E−14	0.9132	0.24840	0.10617
		100	3.34E−13	0.7469	0.16604	0.06018

(continued)

Table 6.1 (continued)

Functions	Rep. method	Dim	Best	Worst	σ	Average
Hyper	Pollination	30	1.8E−158	7.49E−51	1.42E−51	3.55E−52
		50	1.0E−137	8.48E−36	1.62E−36	3.86E−37
		100	2.8E−121	1.61E−23	4.04E−24	1.08E−24
	Clone	30	0.0002	0.0095	0.002113	0.00445
		50	0.0235	0.9688	0.309161	0.51737
		100	0.0145	1.9986	0.608632	1.06511
	Graft	30	3.67E−91	1.3364	0.387834	0.30489
		50	5.6E−106	9.2512	2.159921	1.06361
		100	1.13E−38	38.9208	9.614367	4.64223
Levy	Pollination	30	0.001473	1.36153	0.343047	0.11711
		50	0.008944	1.18304	0.304007	0.14539
		100	0.062205	8.90276	2.065917	1.11298
	Clone	30	0.189020	3.23484	0.843312	0.76403
		50	1.67E−05	0.94790	0.297364	0.31374
		100	0.03603	0.99425	0.407219	0.49375
	Graft	30	0.13535	3.25722	1.100443	0.95391
		50	0.30798	3.81686	1.502540	1.75416
		100	0.60231	6.88429	1.888266	2.20147
Dixon	Pollination	30	0.6666	1.58570	0.167768	0.69743
		50	0.6666	4.88802	0.770709	0.80738
		100	0.00481	0.66671	0.121197	0.64198
	Clone	30	0.5160	6.05080	0.972015	1.25908
		50	0.0113	0.9819	0.296911	0.41073
		100	0.0284	1.1959	0.324379	0.51037
	Graft	30	0.2898	5.9931	1.433557	1.52608
		50	0.1856	18.2151	4.847751	3.00834
		100	0.96215	30.8223	5.678660	3.36568

case study shown in this book, giving the following results shown in Table 6.2. In applying the statistical z test, with a confidence level of 95%, and the alternative hypothesis states that the average of the method of reproduction by pollination is lower than the average of the method of reproduction by graft, and of course the null hypothesis tells us that the average of the method of reproduction by pollination is greater than or equal to, the average of the method of reproduction by graft, with a rejection region for all values that fall below level of −1.645. With a Z value of **−20.696** we can conclude that the pollination reproductions method is more efficient than the method of reproduction by graft. For the function of the sphere and in Table 6.3, the statistical results for all the functions used in this case are shown.

Table 6.2 Parameters for the statistical Z test

Parameters	Values
Confidence level	95%
Alpha	0.05
Ha	$\mu1 < \mu2$
H0	$\mu1 \geq \mu2$
Critical value	−1.645

Table 6.3 Results for the statistical Z test

Function name	Reproduction method		Z-value	Evidence
Ackley	Pollinations	Graft	−5.7692	Significant
Rastrigin			−6.3908	Significant
Sphere			−20.696	Significant
Griewank			−3.3290	Significant
Powell			−5.0589	Significant
Hyper			−12.187	Significant
Levy			−32.796	Significant
Dixon			−32.480	Significant
Zakharov			−0.4170	Not significant
Sum Square			−19.105	Significant
Rosenbronck			−0.3872	Not significant

Analyzing the results shown with statistical test, we can notice that the method of reproduction by pollination is more efficient compared with the others, for this problem, however the other proposed methods on some number of iterations found many values near to the minimum values of the function, and therefore are efficient, but not the best for this problem.

In the previous statistical test we can find that for the three proposed reproduction methods in this book, the best so far is, reproduction by pollination using Levy flights.

6.2   Case Study Two: Optimization of CEC-2015 Benchmark Mathematical Functions

This section shows the results obtained from the experiments performed using the optimization algorithm bio-inspired on the self-defense mechanisms of plants to the set of eight functions from the CEC-2015 competition [3, 4]. Based on previous publications the authors recommend using the method of pollination as a reproduction operator, because it has a higher performance. In this test, 30 experiments

Table 6.4 Mathematical functions

Type	No.	Function
Unimodal functions	F1	Rotated High Conditioned Elliptic Function
	F2	Rotated Cigar Function
Simple multimodal functions	F3	Shifted and Rotated Ackley's Function
	F4	Shifted and Rotated Rastrigin's Function
	F5	Shifted and Rotated Schwefel's Function
Hybrid functions	F6	Hybrid Function 1 ($N = 3$)
	F7	Hybrid Function 2 ($N = 4$)
	F8	Hybrid Function 3 ($N = 5$)

were performed for the following mathematical functions and the evaluation is for 10, 30 Variables, some data of the used functions can be find in Table 6.4, but for more information of the functions please review [3].

6.2.1   Parameters for the Algorithm

For this work the parameters for the variables (α, β, δ, λ), were moved in a specific range, as mentioned before some publications of the algorithm where the authors recommend a range of optimum values to improve the performance of the meta-heuristic [5, 6]. Also the configuration of other parameters such as the size of populations of prey (plants), predators (herbivores) is based on these previous works.

For the CEC 2015 function problems, a range of values are recommended to be able to compete against the results found by other algorithms, but in this work we only want to show that the proposed algorithm can also be used to optimize complex problems. The configuration of parameters are defined below: we use plants = 400, Herbivores = 350, and the ranges for the iterations were 1000–900 to observe the behavior.

6.2.2   Simulation Results

In Table 6.5 we can find the results obtained for the case study used. In Tables 6.5 and 6.6 we observed the results of 30 experiments for each function, using 10 and 30 dimensions; we consider important to the reader the following the worse, best, average, and standard deviation [6].

In Tables 6.5 and 6.6, we show the results obtained from 30 experiments performed for 10 and 30 variables, and we can observe that in the experiments it was very difficult to approximate the value of the function to zero. These mathematical

Table 6.5 Results for 10 dimensions

Function	Important results of the algorithm			
	Best	Worse	σ	Average
F1	4.95E+04	4.38E+06	1.06E+06	1.03E+06
F2	1.40E+05	2.87E+06	7.45E+05	1.15E+06
F3	2.00E+01	2.04E+01	1.08E−01	2.03E+01
F4	8.08E+00	6.67E+01	1.67E+01	2.69E+01
F5	2.45E+02	1.08E+03	2.07E+02	6.24E+02
F6	3.55E+02	4.75E+04	8.66E+03	5.93E+03
F7	1.42E+00	1.23E+01	1.96E+00	2.89E+00
F8	9.41E+02	6.75E+03	1.34E+03	2.30E+03

Table 6.6 Results for 30 dimensions

Function	Important results of the algorithm			
	Best	Worse	σ	Average
F1	2.80E+06	2.94E+07	6.77E+06	1.199E+07
F2	1.74E+07	7.07E+09	1.34E+09	4.275E+08
F3	2.02E+01	2.10E+01	1.58E−01	2.09E+01
F4	1.62E+02	2.99E+02	3.90E+01	2.132E+02
F5	2.67E+03	5.54E+03	7.77E+02	3.91E+03
F6	3.57E+02	4.86E+04	8.82E+03	5.14E+03
F8	2.67E+04	1.18E+06	2.29E+05	2.22E+05

functions used very complex, some are hybrid, multimodal and composite, and this increases the complexity therefore the algorithms have to be more efficient to be able to solve those functions or use the help of other intelligent techniques such as fuzzy logic or the hybridization with another optimization algorithm. It is important to mention that some of the functions do not have their objective value as zero.

6.2.3 Statistical Comparison

To conclude this case study, it is necessary to make a statistical comparison against other published results, and the test used is the z-test. In Table 6.7 we can observe the parameters used in this test, the results obtained with the algorithm of the mechanisms of the plants (MSPA) are compared with Iterative hybridization of differential evolution (DE) with Local Search for the CEC'2015 Special Section on Large Scale Global Optimization (IHDELS) [4].

In applying the statistic Z-test, with a confidence level of 95%, and the alternative hypothesis says that the average of the proposed method is lower than the

Table 6.7 Parameters for statistical comparison

Parameters	Values
Confidence level	95%
Alpha	0.05
Ha	$\mu1 < \mu2$
H0	$\mu1 \geq \mu2$
Critical value	−1.645

Table 6.8 Results of applying the statistical Z test for 30D

Case study	Our method	DynFWA	Z-value	Evidence
F1	MSPA	IHDELS	9.521	Not significant
F2			1.7474	Not significant
F3			19.8598	Not significant
F4			−13.1429	Significant
F5			−23.5751	Significant
F6			−43.48	Significant
F7			−14.8355	Significant

The table presents the statistical results of the proposed method, where the success is observed in some functions presented in comparison with the algorithm of differential evolution (DE) [4]

average of IHDELS [4], and of course the null hypothesis tells us that the average of the proposed method is greater than or equal to the average of IHDELS [4], with a rejection region for all values fall below of −1.645. In Table 6.8 we can observe the results of the statistical comparison.

6.3 Case Study Three: Proposed Algorithm with Dynamic Adjustment Using Type-1 Fuzzy Logic Approach

In this study case the authors propose a variant of the original algorithm of the plants with a type 1 fuzzy logic approach. The new proposal consists of adding fuzzy logic to dynamically adapt the parameters of the algorithm. In this case, a fuzzy controller is responsible of finding the optimal values of the parameters α, β, δ, λ, in order to help the algorithm to have a better performance in solving problems. In the previous works the authors apply the original algorithm to optimization problems, and the parameters of the variables are moved manually, however the results obtained are acceptable in some cases, but we consider that they can be improved using an intelligent technique for the adaptation of parameters.

The proposed new variant is used to optimize a set of benchmark mathematical functions CEC 2015 [3, 4], the functions are shown in Table 6.9. The authors

Table 6.9 Mathematical functions

Type	No.	Function
Unimodal functions	F1	Rotated High Conditioned Elliptic Function
	F2	Rotated Cigar Function
Simple multimodal functions	F3	Shifted and Rotated Ackley's Function
	F4	Shifted and Rotated Rastrigin's Function
	F5	Shifted and Rotated Schwefel's Function

propose to use fuzzy logic to automatically find the best values of the parameters of the model of prey predator. Fuzzy logic is a novel technique to help algorithms solve complex optimization problems [7–10].

The goal of using fuzzy logic in the algorithm is to improve its exploration and use the exploitation at appropriate times, the fuzzy logic controller used in this work is of type Mamdani [11, 12], with 2 input variables and 4 output variables. In Fig. 6.1 we can observe the characteristics of the fuzzy logic controller used in this work.

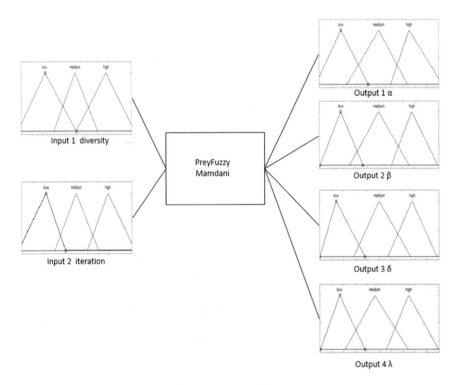

Fig. 6.1 Fuzzy controller proposed for parameter adaptation

Figure 6.1 shows the characteristics of the controller used. The controller has two inputs, with three Gaussian type membership functions, granulation is: low, medium and high, in a range of 0–1. The controller used has four outputs of triangular type, granulated in three membership functions, high, medium and low. This controller is used to dynamically change the values of the parameters (α, β, δ, λ). These parameters represent the birth rate and mortality of the plants and the birth rate of predators and the amount of depredation.

6.3.1 Simulation Result

The proposal with fuzzy approach is tested with functions CEC2015 [3, 4] for 2 and 10 variables, these functions have a high level of complexity, The name of the functions and the characteristics of the controller used in this test is shown in Fig. 6.1. In Table 6.10 we consider it important to show the following data, size of the population of plants (prey) = 300, predators = 250, number of iterations = 500. In Tables 6.10 and 6.11 we can observe the 30 experiments performed for the set of functions. The authors consider it important to show the following data obtained from the simulations, the best, the worst, standard deviation and average for each function evaluated.

In Table 6.10, we can observe the performance of the algorithm using fuzzy logic, and we can observe that the results improved considerably. Our proposal achieves acceptable results in functions F3 and F5. In Table 6.11 we can observe the results obtained for 10 dimensions, where the best results were obtained in the functions F3, F4 and F5. This algorithm is recent however we have achieved favorable results, we consider that these results can be improved considering the following: Perform more experiments and change the type of membership function of the controller, change the reproduction operators, use other combinations of fuzzy rules in the controller and test the algorithm in other optimization problems such as: Optimization of neural networks, fuzzy logic for mention some and observe the behavior.

Table 6.10 Results for 2 dimensions

Function	Results							
	α	β	λ	δ	Best	Worse	σ	Average
F1	Dynamic				9.07E−02	2.36E+01	6.40E+00	6.280E+00
F2	Dynamic				9.07E−02	2.36E+01	6.40E+00	6.280E+00
F3	Dynamic				1.10E−02	1.81E−01	4.94E−02	7.94E−02
F4	Dynamic				8.42E−06	3.22E−03	8.77E−04	7.858E−04
F5	Dynamic				1.13E−04	8.14E−02	2.25E−02	1.90E−02

Table 6.11 Results for 10 dimensions

Function	Important results of the algorithm							
	α	β	λ	δ	Best	Worse	σ	Average
F1	Dynamic				1.72E+06	1.04E+07	2.16E+06	5.87E+06
F2	Dynamic				3.74E+08	2.23E+09	4.28E+08	1.20E+09
F3	Dynamic				2.01E+01	2.05E+01	1.00E−01	2.03E+01
F4	Dynamic				2.50E+01	4.95E+01	6.67E+00	4.01E+01
F5	Dynamic				5.66E+02	1.11E+03	1.31E+02	9.11E+02

6.3.2   Preliminary Conclusions

To conclude this study case we consider important to mention that the use of intelligent techniques in bio-inspired algorithms has had a great success in recent years, and we consider it important to use it in the proposed algorithm and observe the behavior, where our main contribution is the use of fuzzy logic to automatically adjust the values of the parameters (α, β, δ, λ) that are responsible for maintaining a balance between the two populations, controlling the percentage of birth and mortality of the prey and predators, and other important data. It is important to emphasize that the use of fuzzy logic in this case considerably improves the performance of the method, achieving favorable results. This algorithm is recent however we have achieved favorable results, we consider that these results can be improved considering the following data: Perform more experiments and change the type of membership function and change the combination of fuzzy rules of the fuzzy controller, change the reproduction operators, for mention some.

6.4   Case Study Three: Proposed Algorithm with Dynamic Reproduction Method Using Type 1 Fuzzy Approach

In this case a meta-heuristics of optimization based on the mechanisms of self-defense of the plants with dynamic adaptation of parameters using fuzzy logic is presented. In this section we propose a new variant of the algorithm with a type 1 fuzzy approach. The function of fuzzy logic in this work is the dynamic adaptation of the values of the variables (α, β, λ, δ), and also to find the optimal method of biological reproduction for the problem. In this case optimization of the CEC-2015 functions, the proposed algorithm has three operators of biological reproduction (**clone, graft and pollination**), the authors considered those the most common in the nature, as mentioned in Chap. 2. Figure 6.2 describes the steps of the optimization algorithm with the fuzzy approach.

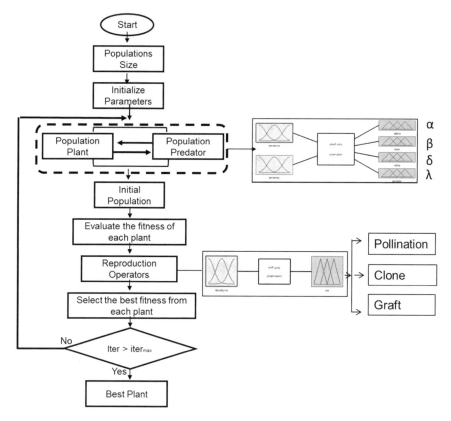

Fig. 6.2 Flowchart with the type 1 fuzzy approach

As mentioned above, this fuzzy controller can select one of the methods of biological reproduction proposed for this test, and the operators are as follows [5].

Clone: The offspring are identical to the parent plant, in this case we use as the basis of the plant that survived along the iterations, and this plant is used to inherit their characteristics to future generations.

Graft: In our proposal a basis a plant with better fitness is used as it takes a stem of a plant and is encrusted on another to generate an alteration in the structure of the plant.

Pollination: Pollination is a biological reproduction method used by plants to send grains of pollen from one plant (flower) to another plant (flower). In order for this process to be performed it depends on several factors.

6.4.1   The Fuzzy Controller Number Two Has the Following Characteristics

This fuzzy system has an input variable (iteration), granulated in three membership functions of Gaussian type (low, medium, high), distributed in a range of [0–1]. It has an output variable called mr (reproduction method), granulated into three membership functions of triangular type (mr1, mr2, mr3) distributed in a range of [0–1]. In Fig. 6.3 we can find an illustration of the variables of the fuzzy controller.

The performance of the new proposed variant with the fuzzy approach is tested in mathematical benchmark functions of CEC-2015 [3, 4], the algorithm is tested for 10 and 30 dimensions, for a set of functions shown in Table 6.12. The configuration of parameters of the algorithm is as follows: iterations 200, population of prey is 300, and predators is 250, values of $(\alpha, \beta, \lambda, \delta)$ are determined by the fuzzy controller number one, the reproduction operator is determined using fuzzy controller number two. The results obtained from the simulations performed for 10 variables are shown in Table 6.13, and Table 6.14 shows the results for 30 variables. In the tables we only consider the most relevant data of the simulations (best worst, average and standard deviation).

Analyzing the results shown in Tables 6.13 and 6.14, we can observe that the algorithm has had significant improvements in comparison with other results published [4], some of the functions selected in this work have a level of complexity that is high, however our proposal have achieved success in the F3, F4, F5, F7 functions in 10 dimensions. For 30 dimensions they were successful in F3, F4, F7, in the other functions the success achieved is not as significant as in the mentioned functions.

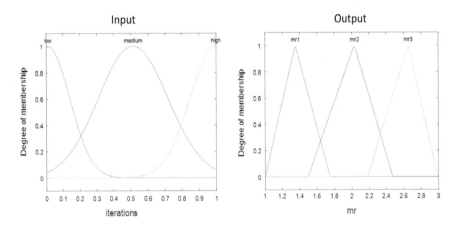

Fig. 6.3 Variables of the fuzzy system

Table 6.12 Mathematical functions

Type	No.	Function name
Unimodal functions	1	Rotated High Conditioned Elliptic Function
	2	Rotated Cigar Function
Simple multimodal functions	3	Shifted and Rotated Ackley's Function
	4	Shifted and Rotated Rastrigin's Function
	5	Shifted and Rotated Schwefel's Function
Hybrid functions	6	Hybrid Function 1 ($N = 3$)
	7	Hybrid Function 2 ($N = 4$)
	8	Hybrid Function 3 ($N = 5$)

Table 6.13 Simulation for 10 dimensions

Function	Results								
	α	β	λ	δ	M. Rep	Best	Worst	σ	Average
F1	Dyn				Dyn	3.57E+05	1.84E+06	3.73E+05	1.12E+06
F2	Dyn				Dyn	2.47E+07	7.84E+07	1.38E+07	5.14E+07
F3	Dyn				Dyn	1.65E+01	2.01E+01	9.68E−01	1.96E+01
F4	Dyn				Dyn	1.16E+01	2.11E+01	2.61E+00	1.55E+01
F5	Dyn				Dyn	1.40E+02	5.10E+02	7.70E+01	3.70E+02
F6	Dyn				Dyn	1.38E+03	1.51E+04	2.86E+03	4.95E+03
F7	Dyn				Dyn	2.32E+00	3.21E+00	2.29E−01	2.75E+00
F8	Dyn				Dyn	4.19E+02	3.04E+03	6.00E+02	1.17E+03

Table 6.14 Simulation for 30 dimensions

Function	Results								
	α	β	λ	δ	M. Rep	Best	Worse	σ	Average
F1	Dyn				Dyn	2.42E+07	6.01E+07	9.22E+06	4.418E+07
F2	Dyn				Dyn	3.08E+09	7.76E+09	1.01E+09	5.838E+09
F3	Dyn				Dyn	2.03E+01	2.05E+01	4.51E−02	2.055E+01
F4	Dyn				Dyn	1.28E+02	1.68E+02	1.13E+01	1.507E+02
F5	Dyn				Dyn	4.11E+03	4.88E+03	1.94E+02	4.544E+03
F6	Dyn				Dyn	6.78E+05	3.64E+06	7.33E+05	2.05E+06
F7	Dyn				Dyn	2.32E+00	3.21E+00	2.29E−01	2.75E+00
F8	Dyn				Dyn	1.35E+05	7.53E+05	1.38E+05	4.98E+05

6.4.2 Preliminary Conclusions

We have observed that the use of fuzzy logic in combination with optimization algorithms significantly improves results and stability. Analyzing the results we observed that the performance is acceptable, but we consider that these results could be improved by making some improvements in the fuzzy controllers, for example changing the type of the membership functions in the input and output variables and perform more tests and compare the results. It is also necessary to apply our proposed meta-heuristics to other optimization problems and test the performance of the algorithm and compare it with other meta-heuristics, it is important to mention that this meta-heuristic is relatively recent.

6.5 Case Study Three: Proposed Algorithm with a Type-2 Fuzzy Approach

In this section, a new variant of the algorithm bio-inspired in the self-defense mechanisms of plants in nature with dynamic adaptation of parameters using Type-2 FLS applied to the optimization of the trajectory of a mobile robot is presented. There are multiple works using Type-2 FLS applied to different optimization problems, the use of this technique significantly improves the results, like in [5, 13]. The principles of Type-2 FLS can be consulted in [14–16]. We decided to combine type-2 fuzzy logic with our proposal based on works found in the literature, where it is shown that type 2 fuzzy controllers offer a higher performance when applied to robust problems. It is important to mention that in this work, our main contribution is the integration of Type-2 fuzzy logic to the proposed algorithm.

In Fig. 6.4, a flowchart representing the process of the algorithm and we can also observe the stage of the algorithm where we apply Type-2 FL to adjust the values of $(\alpha, \beta, \delta, \lambda)$.

For this book we are proposing the method of reproduction by pollination, as in previous works using this algorithm with this method of reproduction the results are acceptable, and in the previous Sect. 6.4 we explain the method of reproduction by pollination.

6.5.1 Characteristics of the Used Fuzzy Controller

The proposed Type-2 FLS controller is of Mamdani Type, this controller has two input variables (iteration and diversity) and four output variables (Alpha, beta, delta, and lambda). In Figs. 6.5 and 6.6, we can observe the characteristics of the

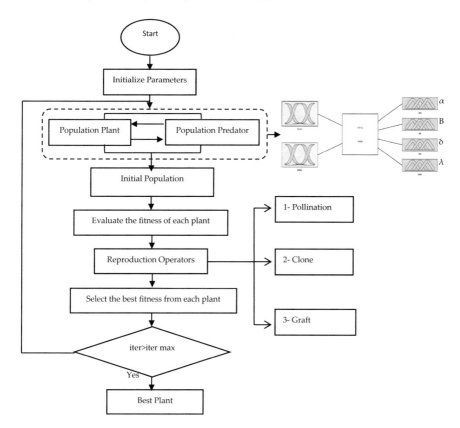

Fig. 6.4 Flowchart with a type-2 fuzzy approach

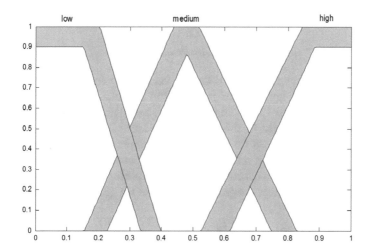

Fig. 6.5 Input variable iteration

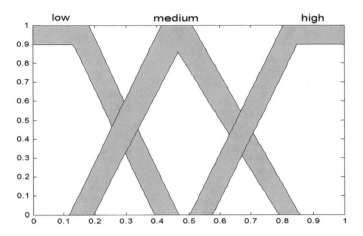

Fig. 6.6 Input variable diversity

input variables. The variable iteration and diversity are granulated into three membership functions (low, medium, and high) in a range of [0–1], the type of membership functions proposed for this work are: triangular in the center and trapezoid in the extremes.

We consider it important to use the iterations value as an input variable of the proposed Type-2 fuzzy controller, using this input variable we can maintain a balance between the exploration and exploitation of the algorithm, because when the iterations are Low we need the algorithm to explore and when the iterations are High we can order it to converge in the best region found.

The "Iteration" variable is defined by Eq. (6.11), and has a range from 0 to 1; this variable can be viewed as the percentage of the current iterations. In Fig. 6.6, the second input variable diversity is presented.

$$Iteration = \frac{Current\ Iteration}{Maximun\ of\ Iteration} \tag{6.11}$$

For the second input variable of the proposed fuzzy controller, we are considering the diversity. The diversity is a metric that helps us measure how much similarity exists between the population and the best plant found so far. The metric used to measure diversity in this work is the Jaccard index, proposed by Paul Jaccard, it is a statistic used for comparing the similarity and diversity of sample sets. The Jaccard coefficient measures similarity between finite sample sets, and is defined as the size of the intersection divided by the size of the union of the sample sets. The mathematical representation is shown in Eq. (6.12) [17–20].

$$J(X,Y) = \frac{|X \cap Y|}{|X \cup Y|} \tag{6.12}$$

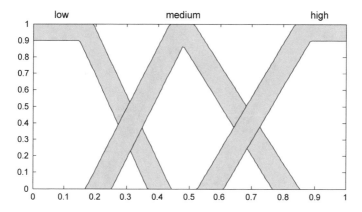

Fig. 6.7 Output 1 Alpha

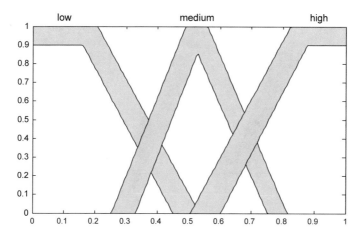

Fig. 6.8 Output 2 Beta

where:

Count the number of members which are similar between both populations.
Count the total number of members in both populations. (Similar and not-similar).
Divide the number of similar members (1) by the total number of members (2).
Multiply the number we found in (3) by 100.

This metric has also been used by other authors in their works, for more details consult in [3, 4, 8, 21]. In Figs. 6.7, 6.8, 6.9 and 6.10 the four output variables are shown.

The output variables are granulated into three membership functions (Low, Medium, and High) with a range of [0–1], with triangular membership functions in the center and trapezoid in the extremes, and all variables have the same configuration.

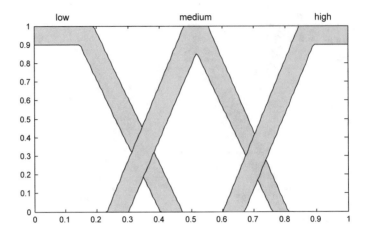

Fig. 6.9 Output 3 Delta

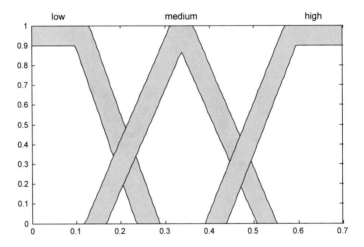

Fig. 6.10 Output 4 Lambda

It is important to mention that these configurations were selected based on prior knowledge of experiments performed and this configuration was the one that gave the best result.

6.5.2 Characteristics of the Problem to Optimize

The case selected to apply the proposed algorithm with Type-2 fuzzy approach is the problem of the autonomous robot [25, 26]. The dynamics of this problem is that

Fig. 6.11 Representation of the mobile robot

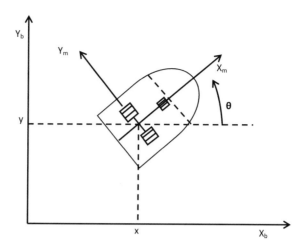

the robot has to be able to follow an assigned trajectory with a minimum error value. In the proposed method an FLS is used to move the parameters of the membership functions of the fuzzy controller. The description of the problem and the characteristics of the fuzzy controller are shown below.

Is an autonomous vehicle capable of following predictable paths in uncertain environments and is illustrated in Fig. 6.11. The robot body is symmetrical around the perpendicular axis and the center of mass is at the geometric center of the body.

It has two driving wheels that are fixed to the axis that passes through the center of mass "C" represented by {C, Xm, Ym}, and one passive wheel that prevents the robot from tipping over as it moves on a plane [22–24]. The dynamics of the mobile robot is represented by the following set of Eqs. (6.13) and (6.14) [25, 26].

$$M(q)\dot{v} + C(q,\dot{q})v + Dv = \tau + P(t) \tag{6.13}$$

where,

$q = (x, y, \theta)^T$ is the vector of the configuration coordinates,
$\upsilon = (v, w)^T$ is the vector of velocities,
$\tau = (\tau_1, \tau_2)$ is the vector of torques applied to the wheels of the robot where τ_1 and τ_2 denote the torques of the right and left wheel,
$P \in R^2$ is the uniformly bounded disturbance vector,
$M(q) \in R^{2\times2}$ is the positive-definite inertia matrix,
$C(q,\dot{q})\vartheta$ is the vector of centripetal and Coriolis forces, and
$D \in R^{2\times2}$ is a diagonal positive-definite damping matrix.

The kinematic system is represented by Eq. (6.14)

$$\dot{q} = \underbrace{\begin{bmatrix} \cos\theta & 0 \\ \sin\theta & 0 \\ 0 & 1 \end{bmatrix}}_{J(q)} \underbrace{\begin{bmatrix} v \\ w \end{bmatrix}}_{v} \tag{6.14}$$

where:

(x, y) is the position in the X–Y (world) reference frame,
θ is the angle between the heading direction and the x-axis, and
v and w are the linear and angular velocities.

Furthermore Eq. (6.15) shows the non-holonomic constraint, which this system has, which corresponds to a no-slip wheel condition preventing the robot from moving sideways.

$$\dot{y}\cos\theta - \dot{x}\sin\theta = 0 \tag{6.15}$$

The system fails to meet Brockett's necessary condition for feedback stabilization, which implies that no continuous static state-feedback controller exists that can stabilize the closed-loop system around the equilibrium point.

6.5.3 Characteristics of the Fuzzy Controller Used for the Robot

The characteristics of the fuzzy controller used in this case study are presented below. We chose to optimize a fuzzy controller of the path for a unicycle mobile robot to test the proposed method in a more complex problem. The controller is of Mamdani type, so that the input and output parameters are represented by linguistic variables. The input variables are the error in the linear velocity (ev) and angular velocity (ew), and the output variables are the right (T1) and left (T2) torques [25, 26], which are represented in Fig. 6.12.

The membership functions of the input variables are of trapezoidal type in the negative (N) and positive (P) linguistic terms and triangular for the linguistic term for zero (Z). For the output variables we have three membership functions, negative (N), zero (Z), and positive (P) of triangular type. In the range of values of $[-1, 1]$ is used for each variable because we have normalized all the variables.

The fuzzy rules that are used are shown below

If (ev is N) and (ew is N) then (T1 is N) (T2 is N).
If (ev is N) and (ew is Z) then (T1 is N) (T2 is Z).
If (ev is N) and (ew is P) then (T1 is N) (T2 is P).

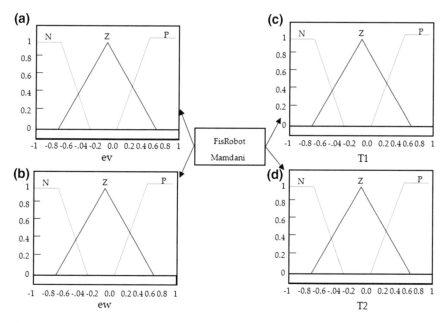

Fig. 6.12 a Linear velocity error (ev); **b** angular velocity error (ew); **c** left torque (T1); **d** right torque (T2)

If (ev is Z) and (ew is N) then (T1 is Z) (T2 is N).
If (ev is Z) and (ew is Z) then (T1 is Z) (T2 is Z).
If (ev is Z) and (ew is P) then (T1 is Z) (T2 is P).
If (ev is P) and (ew is N) then (T1 is P) (T2 is N).
If (ev is P) and (ew is Z) then (T1 is P) (T2 is Z).
If (ev is P) and (ew is P) then (T1 is P) (T2 is P).

The trajectory to be traversed by the robot is shown in Fig. 6.13.

The path given to the robot is constant (does not change), and in the course of iterations to the algorithm in combination of Type-2 FLS are responsible for adapting the parameters to direct the robot in that trajectory.

6.5.4   Simulation Results

The proposed algorithm using Type-2 FLS to adapt the values of the variables of the predator prey model is used to optimize the trajectory of the problem of the mobile autonomous robot, with the purpose of measuring their performance and stability in complex problems. It is important to emphasize that the proposed algorithm takes as a basis the functions of the predator-prey model. In Table 6.15, we can observe the configuration of parameters used in this case study to perform the experiments.

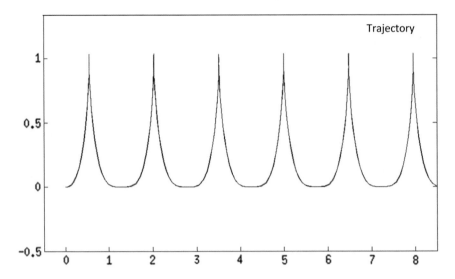

Fig. 6.13 Trajectory to be followed by robot

The parameters shown in Table 6.15 were selected based on previous results and also are used in other works where the same algorithm is used, but now applied to a different problem. Also, the results of 30 experiments performed are shown, analyzing the results we observed that the algorithm with Type-2 fuzzy approach has a higher performance and stability. In the values we can observe that they are very similar to the others, that tells us that the algorithm is stable since all experiments are successful, finding acceptable values and we can affirm that the robot is able to follow a trajectory with a greater performance. In Fig. 6.14, the best robot simulation following the reference is presented.

In Fig. 6.14, we can observe a better performance of the fuzzy controller built by the proposed algorithm using Type-2 FLS (see experiment number 9 in Table 6.15), the blue line represents the reference and the green line represents the trajectory traveled by the robot. The proposed approach has greater stability and adaptation exceeding the expectations of the authors. The errors shown in Table 6.15 were calculated using the Eq. (6.16) Mean square error [6, 27].

$$MSE = \frac{1}{N}\sum_{K=1}^{N}[x(k) - y(k)]^2 \qquad (6.16)$$

In Eq. (6.16), x(k) represents the actual reference value at time k, y(k) represents the output value of the system at time k, and N represents the total number of samples considered.

Table 6.15 Results obtained and parameters of the algorithm

N	Iter	α	β	δ	λ	Prey	Predator	Values
		Variables				Populations		
1	50	Dynamic with Interval				300	200	7.67×10^{-2}
2	50	Type-2 FLS				300	200	1.76×10^{-2}
3	50					300	200	1.12×10^{-2}
4	50					300	200	7.76×10^{-2}
5	50					300	200	7.67×10^{-2}
6	50					300	200	1.54×10^{-2}
7	50					300	200	1.54×10^{-2}
8	50					300	200	9.62×10^{-3}
9	50					300	200	8.25×10^{-5}
10	50					300	200	2.08×10^{-3}
11	50					300	200	1.77×10^{-3}
12	50					300	200	2.17×10^{-2}
13	50					300	200	3.09×10^{-2}
14	50					300	200	3.38×10^{-2}
15	50					300	200	1.05×10^{-2}
16	50					300	200	1.69×10^{-2}
17	50					300	200	3.54×10^{-2}
18	50					300	200	2.79×10^{-2}
19	50					300	200	5.22×10^{-3}
20	50					300	200	1.56×10^{-2}
21	50					300	200	1.25×10^{-2}
22	50					300	200	3.82×10^{-3}
23	50					300	200	1.82×10^{-2}
24	50					300	200	2.27×10^{-3}
25	50					300	200	7.27×10^{-3}
26	50					300	200	3.70×10^{-3}
27	50					300	200	6.93×10^{-4}
28	50					300	200	2.02×10^{-2}
29	50					300	200	4.25×10^{-3}
30	50					300	200	6.93×10^{-4}
						Average		1.919×10^{-2}
						σ		2.20×10^{-2}
						Best		8.25×10^{-5}
						Worst		7.76×10^{-2}

6.5.5   Statistical Comparison

To conclude this work, we consider it important to make a statistical comparison to verify if there is scientific evidence that the algorithm with a Type-2 FLS approach

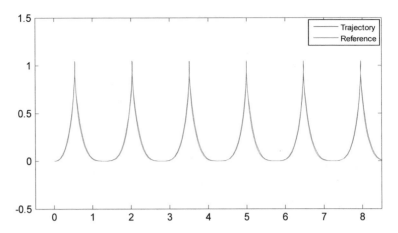

Fig. 6.14 Robot simulation

Table 6.16 Statistical z-test parameters

Parameters	Values
Confidence level	95%
Alpha	0.05
Ha	$\mu1 < \mu2$
H0	$\mu1 \geq \mu2$
Critical value	-1.645

is successful in this case study compared to other works published by other authors and show that our proposal offers a significant contribution. The statistical comparison is made against the Bee Colony Optimization Algorithm with fuzzy logic (FBCO) [6, 23, 25]. The statistical test used for the comparisons is the z-test, whose parameters are defined in Table 6.16.

A confidence level of 95% is used (that is a usual value), the alternative hypothesis states that the average of the self-defense of the plant algorithm is lower than the average of FBCO, and the null hypothesis states that the average of the proposed algorithm is greater than or equal to the average of FBCO, with a rejection region for all values falling below -1.645. The statistical test results are that: for the self-defense of the plant algorithm with a fuzzy approach, there is significant evidence to reject the null hypothesis with a calculated Z value of -3.9163. Clearly this value falls in the rejection region, therefore, we have enough statistical evidence to affirm that the algorithm proposed is less than the bee colony algorithm with a diffuse approach.

Analyzing the results of the statistical test we can affirm that we have enough evidence to conclude that the average of our proposal is better than the bee colony algorithm with a fuzzy approach, there is also enough statistical evidence to affirm that in this book we have a successful contribution, and we also observed that the use of FLS as a complement in optimization algorithms significantly improve performance and stability in optimization problems in comparison with other previous works published by the same author of the proposed algorithm [6].

References

1. Yang, X. S. (2013). Multiobjective firefly algorithm for continuous optimization. *Engineering with Computers, 29*(2), 175–184.
2. Yang, X. S. (2010). Firefly algorithm, stochastic test functions and design optimisation. *International Journal of Bio-Inspired Computation, 2*(2), 78–84.
3. Liang, J. J., Qu, B. Y., Suganthan, P. N., & Chen, Q. (2014). *Problem definitions and evaluation criteria for the CEC 2015 competition on learning-based real-parameter single objective optimization.* Technical Report 201411A, Computational Intelligence Laboratory, Zhengzhou University, Zhengzhou China and Technical Report, Nanyang Technological University, Singapore.
4. Molina, D., & Herrera, F. (2015, May). Iterative hybridization of DE with local search for the CEC'2015 special session on large scale global optimization. In *2015 IEEE Congress on Evolutionary Computation (CEC)* (pp. 1974–1978). IEEE.
5. Caraveo, C., Valdez, F., & Castillo, O. (2015). A new bio-inspired optimization algorithm based on the self-defense mechanisms of plants. In *Design of intelligent systems based on fuzzy logic, neural networks and nature-inspired optimization* (pp. 211–218). Springer International Publishing.
6. Caraveo, C., Valdez, F., & Castillo, O. (2015). Bio-inspired optimization algorithm based on the self-defense mechanism in plants. In *Advances in artificial intelligence and soft computing* (pp. 227–237). Springer International Publishing.
7. Johanyák, Z. C., & Papp, O. (2012). A hybrid algorithm for parameter tuning in fuzzy model identification. *Acta Polytechnica Hungarica, 9*(6), 153–165.
8. Melin, P., Castillo, O., Gonzalez, C. I., Castro, J. R., & Mendoza, O. (2016, October). General Type-2 fuzzy edge detectors applied to face recognition systems. In *Fuzzy Information Processing Society (NAFIPS), 2016 Annual Conference of the North American* (pp. 1–6). IEEE.
9. Ochoa, P., Castillo, O., & Soria, J. (2016, September). Fuzzy differential evolution method with dynamic parameter adaptation using type-2 fuzzy logic. In *2016 IEEE 8th International Conference on Intelligent Systems (IS)* (pp. 113–118). IEEE.
10. Precup, R. E., David, R. C., Petriu, E. M., Preitl, S., & Rădac, M. B. (2014). Novel adaptive charged system search algorithm for optimal tuning of fuzzy controllers. *Expert Systems with Applications, 41*(4), 1168–1175.
11. Singla, J. (2015, March). Comparative study of Mamdani-type and Sugeno-type fuzzy inference systems for diagnosis of diabetes. In *2015 International Conference on Advances in Computer Engineering and Applications (ICACEA)* (pp. 517–522). IEEE.
12. Zi, B., Sun, H., & Zhang, D. (2017). Design, analysis and control of a winding hybrid-driven cable parallel manipulator. *Robotics and Computer-Integrated Manufacturing, 48,* 196–208.
13. Caraveo, C., Valdez, F., Castillo, O., & Melin, P. (2016, December). A new metaheuristic based on the self-defense techniques of the plants in nature. In *2016 IEEE Symposium Series on Computational Intelligence (SSCI)* (pp. 1–5). IEEE.
14. González, C. I., Castro, J. R., Martínez, G. E., Melin, P., & Castillo, O. (2013, June). A new approach based on generalized type-2 fuzzy logic for edge detection. In *IFSA World Congress and NAFIPS Annual Meeting (IFSA/NAFIPS), 2013 Joint* (pp. 424–429). IEEE.

15. González, C. I., Melin, P., Castro, J. R., Castillo, O., & Mendoza, O. (2016). Optimization of interval type-2 fuzzy systems for image edge detection. *Applied Soft Computing, 47,* 631–643.
16. Olivas, F., Valdez, F., Castillo, O., Gonzalez, C. I., Martinez, G., & Melin, P. (2017). Ant colony optimization with dynamic parameter adaptation based on interval type-2 fuzzy logic systems. *Applied Soft Computing, 53,* 74–87.
17. Barbosa, A. M. (2015). fuzzySim: Applying fuzzy logic to binary similarity indices in ecology. *Methods in Ecology and Evolution, 6*(7), 853–858.
18. Gupta, A. K., & Sardana, N. (2015, August). Significance of clustering coefficient over Jaccard index. In *2015 Eighth International Conference on Contemporary Computing (IC3)* (pp. 463–466). IEEE.
19. Hi, R., Ngan, K. N., & Li, S. (2014, October). Jaccard index compensation for object segmentation evaluation. In *2014 IEEE International Conference on Image Processing (ICIP)* (pp. 4457–4461). IEEE.
20. Ramli, N., & Mohamad, D. (2010, December). Fuzzy evaluation based on Jaccard with degree of optimism ranking index. In *2010 International Conference on Science and Social Research (CSSR)* (pp. 970–974). IEEE.
21. Melin, P., Olivas, F., Castillo, O., Valdez, F., Soria, J., & Valdez, M. (2013). Optimal design of fuzzy classification systems using PSO with dynamic parameter adaptation through fuzzy logic. *Expert Systems with Applications, 40*(8), 3196–3206.
22. Amador-Angulo, L., Mendoza, O., Castro, J. R., Rodríguez-Díaz, A., Melin, P., & Castillo, O. (2016). Fuzzy sets in dynamic adaptation of parameters of a bee colony optimization for controlling the trajectory of an autonomous mobile robot. *Sensors, 16*(9), 1458.
23. Caraveo, C., Valdez, F., & Castillo, O. (2016). Optimization of fuzzy controller design using a new bee colony algorithm with fuzzy dynamic parameter adaptation. *Applied Soft Computing, 43,* 131–142.
24. Kennedy, J. (2011). Particle swarm optimization. In *Encyclopedia of machine learning* (pp. 760–766). USA: Springer.
25. Kıran, M. S., & Fındık, O. (2015). A directed artificial bee colony algorithm. *Applied Soft Computing, 26,* 454–462.
26. Paré, P. W., & Tumlinson, J. H. (1999). Plant volatiles as a defense against insect herbivores. *Plant Physiology, 121*(2), 325–332.
27. Zi, B., Zhu, Z. C., & Du, J. L. (2011). Analysis and control of the cable-supporting system including actuator dynamics. *Control Engineering Practice, 19*(5), 491–501.

Chapter 7
Conclusions

In this work we propose a new optimization meta-heuristic that is bio-inspired on the self-defense mechanisms of plants. This algorithm was created recently and we have successfully achieved the integration of the predator prey model to the optimization algorithm and consequently, we adapted some of the commonly methods most used in natural biological reproduction, in this case the authors considered to use graft, clone, and pollination using the Levy flights method.

The three reproduction methods show acceptable results, and therefore our proposal exceeds the expectations of the creators of the optimization algorithm. The main objective was to create a stable and efficient algorithm that is able to solve different optimization problems, in order to achieve in competing against different existing optimization methods in the literature. We should mention that we found optimal ranges of values for the α, β, λ, δ parameters, for this problem, and also for the mathematical functions of the CEC-2015, the proposal shows an acceptable performance. It is important to mention that we proposed a new variant of the algorithm proposed using Type-2 FLS and based on the statistical comparison against the BCO algorithm [1] we have a lot of statistical evidence to say that the self-defense of the plants algorithm with the fuzzy approach, is stable and efficient in the case study applied. We also consider it advisable to use other methods of biological reproduction already integrated in the algorithm. In the case of this study, we only use the method of reproduction pollination using Levy flights, in this work we do not focus on exploring other parameters of the algorithm, since our main objective is to prove that our proposal can be combined with Type-2 FLS to be able to compete against other traditional algorithms and to affirm that although it is recent it is also efficient.

In this book the main contribution was the creation of a new optimization algorithm bio-inspired on the self-defense mechanisms of the plants in nature, with the integration of the predator-prey model and the development of different methods of biological reproduction as internal operators of the proposed algorithm, also the new proposed variant using type 2 fuzzy logic to optimize the parameters of the algorithm.

C. Caraveo et al., *A New Bio-inspired Optimization Algorithm Based on the Self-defense Mechanism of Plants in Nature*, SpringerBriefs in Computational Intelligence, https://doi.org/10.1007/978-3-030-05551-6_7

We consider important to mention that as future work you can have the following: Test the performance of our new variant in other optimization problems such as: optimize some parameters in neural networks; optimize functions of CEC-2017; optimize other problems in the area of fuzzy controllers just to mention some. Test the algorithm with type 2 fuzzy logic, using other methods of biological reproduction example grafting and cloning, or develop others, also apply it to other control benchmark problems.

Reference

1. Olivas, F., Valdez, F., & Castillo, O. (2015). Dynamic parameter adaptation in ant colony optimization using a fuzzy system for TSP problems. In *IFSA-EUSFLAT* (pp. 765–770).

Bibliography

1. Debbarma, S., Saikia, L. C., & Sinha, N. (2014). Solution to automatic generation control problem using firefly algorithm optimized I λ D μ controller. *ISA Transactions, 53*(2), 358–366.
2. Waser, N. M., Chittka, L., Price, M. V., Williams, N. M., & Ollerton, J. (1996). Generalization in pollination systems, and why it matters. *Ecology, 77*(4), 1043–1060.
3. Yang, X. S., Hosseini, S. S. S., & Gandomi, A. H. (2012). Firefly algorithm for solving non-convex economic dispatch problems with valve loading effect. *Applied Soft Computing, 12*(3), 1180–1186.

Index

Printed in the United States
By Bookmasters